DEDICATION

To those who have the courage to envision a more intelligent, interconnected society where innovation improves all facets of existence. This book is devoted to the dreamers, engineers, and visionaries who see the revolutionary potential of technology and its ability to improve the efficiency, meaning, and smoothness of our daily lives. May we all welcome the unseen intelligence that exists all around us and may the future they envisage come to pass.

To my mentors, friends, and family, whose encouragement and support have enabled me to embark on this adventure. I appreciate your steadfast faith in me.

DISCLAIMER

This book's content is intended solely for general informative purposes. Although every attempt has been taken to guarantee the content's accuracy, the author and publisher make no guarantees or representations about the information's accuracy, completeness, or dependability.

The opinions and viewpoints presented in this book are entirely the author's own and may not represent those of any organizations or entities that are referenced. Any direct, indirect, incidental, or consequential damages arising from the use or misuse of the material in this book are not the author's responsibility.

Before making judgments based on the information provided here, readers are urged to carry out their own research and speak with professionals or experts in the pertinent sectors.

CONTENTS

ACKNOWLEDGMENTS 1

CHAPTER 1 1

Comprehending Invisible Ambient Intelligence 1

    1.1 Concept and Definition 1

    1.2 The Development of Sensing Technologies Over Time 3

    1.3 Ambient Invisible Intelligence's salient characteristics 5

CHAPTER 2 7

AII's Foundational Technologies 7

    2.1 Ultra-Low-Power Sensors and Smart Tags 7

    2.2 Protocols for Wireless Communication 9

    2.3 Edge and Cloud Computing Synergy 11

CHAPTER 3 14

Initial Uses in Retail 14

    3.1 Stock and Inventory Control 14

    3.2 Improving the Client Experience 16

    3.3 Cutting Down on Operating Expenses 18

CHAPTER 4 21

Supply Chain Transformation and Logistics 21

    4.1 Monitoring Perishable Items 21

    4.2 Facilitating Traceability and Transparency 23

    4.3 Simplifying the Management of Warehouses 25

CHAPTER 5 28

AII's Revolution in Healthcare 28

5.1 Tracking Medication and Patient Monitoring................................28

5.2 Optimization of Hospital Resources..................................31

5.3 Assisting the Elderly and Disabled..................................33

CHAPTER 6..37

Conservation and Monitoring of the Environment...............37

6.1 Monitoring and Preserving Wildlife.................................37

6.2 Monitoring of Air and Water Quality...............................40

6.3 Early Warning Systems and Disaster Management.......42

CHAPTER 7..45

Smart Cities and Urban Infrastructure.............................45

7.1 Improving Safety for the Public.......................................45

7.2 Energy and Utilities Optimization...................................47

7.3 Intelligent Transportation and Traffic Management........49

CHAPTER 8..52

Everyday Life and Consumer Technology...........................52

8.1 Wearable technology and smart homes............................52

8.2 Tailored User Interfaces...54

8.3 ALL in Learning and Education......................................57

CHAPTER 9..60

Adoption of AII: Obstacles and Hazards............................60

9.1 Security and Privacy Issues..60

9.2 Barriers Caused by Technology.......................................63

9.3 Implications for Ethics and Society.................................66

CHAPTER 10..71

Ambient Invisible Intelligence's Future Prospects.............71

10.1 Greater Incorporation into Daily Life.. 71

10.2 Patterns Through and After 2027... 76

10.3 Establishing an Ecosystem of Sustainable AII.............................. 79

Glossary... **83**

ABOUT THE AUTHOR..**88**

ACKNOWLEDGMENTS

With great appreciation, I would like to thank everyone who helped to make this book possible.

I want to start by sincerely thanking my family for their unfailing encouragement and support along this journey. My greatest strength has come from your belief in me, even in the most trying moments.

I am also extremely appreciative to my mentors and tech colleagues, whose knowledge, experience, and enthusiasm for innovation have influenced the substance of this book. My investigation into the intricacies of Ambient Invisible Intelligence and its revolutionary effects on society has been greatly aided by your advice.

Thank you to my friends and peers who have read, reviewed, and offered insightful comments on my work. I appreciate your time, commitment, and candid thoughts. Your insights have contributed to the development and improvement of this book to its current state.

Lastly, I want to thank the innumerable scientists, researchers, and innovators whose contributions to artificial intelligence, smart sensing, and related technologies continue to motivate and advance the field. Your pioneering work served as the foundation for this publication.

I want to express my gratitude to everyone who has shared in this trip. This book would not have been possible without your contributions, no matter how modest.

CHAPTER 1

Comprehending Invisible Ambient Intelligence

With the integration of sophisticated sensors, smart tags, and intelligent systems into our surroundings without obvious interference, Ambient Invisible Intelligence (AII) is a revolutionary technological advancement. This chapter explores the fundamentals of AII, including its definition, development across time, and salient characteristics.

1.1 Concept and Definition

A New Paradigm: Ambient Invisible Intelligence (AII)
The term "ambient invisible intelligence" describes the advanced fusion of intelligent systems, ultra-low-cost tags, and smart sensors. In contrast to traditional IoT and AI technologies, which require visible equipment or explicit input, AII works in the background and blends in with its surroundings. Unobtrusive intelligence that improves human experience, streamlines procedures, and permits

well-informed decision-making is its aim.

The Role of Invisibility
1. **Seamless Integration:** AII technology is made to blend in with everyday surroundings so well that people frequently don't even notice it. Because of its invisibility, the technology is guaranteed to enhance rather than interfere with human behavior.
2. **Enhanced User Experience:** AII promotes a seamless user experience by doing away with the need for obvious interfaces or direct interactions, especially in industries like healthcare, retail, and urban planning.

How AII Differs from Traditional IoT and AI Systems
- **IoT Comparison:** Wearable trackers and smart thermostats are examples of devices that are openly networked and managed in traditional IoT. However, AII reduces complexity by using passive, low-cost components, such as battery-free tags.
- **AI Comparative Analysis:** AII integrates localized, embedded AI-driven decision-making with real-time environmental sensing, whereas AI concentrates on

data analysis and prediction.
- **Fusion of Strengths**: AII achieves cost-efficiency and scalability by combining the processing power of AI with the sensing capability of IoT.

1.2 The Development of Sensing Technologies Over Time

Early Sensors and Their Limitations
- **Mechanical Sensors:** Mercury thermometers and mechanical pressure sensors were examples of early sensing technologies that were sluggish, heavy, and necessitated extensive manual calibration.
- **High Costs and Narrow Use Cases:** They were limited to specialized industries such as scientific research or industrial automation due to their high cost and restrictions.

The Transition to Smaller, Cost-Effective Smart Tags
- **The Miniaturization Revolution:** Late 20th-century developments in microelectronics produced small, reasonably priced sensors, opening the door to more useful uses.

- **RFID and Beyond:** A major advancement was made with the introduction of radio-frequency identification (RFID) tags. These passive tags could track items wirelessly and didn't require any onboard power.
- **Low-Power Connectivity:** As technologies such as Bluetooth Low Energy (BLE) advanced, the power and cost requirements were further decreased, allowing for widespread implementation.

Important Turning Points for AII

1. **1980s - Sensor Networks:** Proof-of-concept for scalable networks was obtained through early experiments with connected sensors.
2. **2000s - IoT Proliferation:** Although problems like cost and energy usage remained, the quick adoption of IoT showed the need for connected settings.
3. **2020s - Emergence of AII:** Developments in energy-harvesting methods, machine learning, and nanotechnology led to the emergence of AII as a workable and affordable option.

1.3 Ambient Invisible Intelligence's salient characteristics

Cost-Efficiency and Scalability
- **Ultra-Low-Cost Tags:** By incorporating low-cost, printable tags, AII systems are made available even in environments with limited resources, such as rural logistics or public health tracking.
- **Scalable Networks:** Because AII systems use few resources, they can be implemented on a large scale. AII scales without incurring excessive expenses for tasks like supply chain optimization and municipal traffic monitoring.

Real-Time Tracking and Decision-Making
- **Dynamic Response:** AII allows for real-time system modifications, like rerouting delivery trucks or identifying production irregularities, by continuously gathering and evaluating data.
- **Localized Intelligence:** AII uses edge computing to make judgments in real time within their surroundings, rather than depending entirely on

centralized processing.

Minimal Environmental and Energy Footprints

- **Energy Harvesting:** AII systems frequently include energy-harvesting technologies, which derive power from ambient sources like light, vibrations, or radio waves. This results in minimal environmental and energy footprints.
- **Eco-Friendly Materials:** A lot of AII components are made with recyclable materials and electronic waste reduction in mind.
- **Low Power Consumption:** AII reduces its impact on ecosystems and energy grids by eliminating the need for big batteries and intricate infrastructure.

A revolutionary change in the way technology interacts with human surroundings is represented by ambient invisible intelligence. We can better grasp AII's potential to transform industries and enhance lives in significant and imperceptible ways by comprehending its underlying ideas, historical development, and distinctive characteristics.

CHAPTER 2

AII's Foundational Technologies

The successful implementation of Ambient Invisible Intelligence (AII) depends on a number of advanced technologies cooperating. These fundamental technologies allow AII to function smoothly and sustainably, from ultra-low-power sensors and reliable communication protocols to the cooperation between edge and cloud computing. The fundamental components of AII are examined in this chapter, together with its developments, methods, and ramifications.

2.1 Ultra-Low-Power Sensors and Smart Tags

Summary of Smart Tags and Their Downsizing
Smart tags are little gadgets made to gather, store, and send information about the things they are monitoring or their surroundings. Their transformation from large RFID tags to intricate, tiny parts has been revolutionary:

- **Compact Design:** Thanks to developments in microfabrication and nanotechnology, tags as thin as paper and smaller than a postage stamp can now be made.
- **Versatile Applications:** These tags can be incorporated into a variety of settings, such as medical implants for health monitoring or apparel for inventory control.

Methods of Energy Harvesting for Sustainable Operations

AII is based on energy efficiency, and energy-harvesting technologies have completely changed how sensors and smart tags power themselves:

Sources of Ambient Energy:
- **Light:** Artificial or natural light is captured by solar cells built inside sensors.
- **Vibrations:** Piezoelectric materials generate electricity from mechanical motions.
- Rectennas, also known as rectifying antennas, use ambient radio waves to generate power.
- **Eliminating Batteries:** AII minimizes electrical waste and lowers expenses by utilizing these

renewable energy sources to lessen dependency on conventional batteries.

Advanced Materials Enabling Durability and Efficiency
New materials help smart tags and sensors last and work well in a variety of settings:
- **Flexible Electronics:** Tags can adapt to uneven surfaces thanks to printed circuits on flexible substrates.
- **Self-Healing Materials:** AII devices can last longer thanks to emerging polymers that can fix damage.
- **Eco-Friendly Alternatives:** Materials that support sustainability objectives include recycled parts and biodegradable polymers.

2.2 Protocols for Wireless Communication

The Function of Near-Field Communication (NFC), RFID, and Bluetooth Low Energy (BLE)
The foundation of AII is wireless communication protocols, which enable smooth data transfer between devices:
- **BLE:** Distinguished by its low power consumption,

BLE facilitates continuous data interchange in applications such as fitness trackers and smart home systems.

RFID:
- **Passive RFID:** Uses electromagnetic induction to function and doesn't require any onboard power.
- **Active RFID:** Incorporates a small battery for extended range and functionality.
- **NFC:** Perfect for close-quarters communication, NFC allows for safe data transfer and transactions, including keyless access and mobile payments.

5G and Beyond Integration for Wide-Scale Connectivity

5G network deployment offers AII systems hitherto unheard-of advantages:

- **High Bandwidth:** Supports massive data transfer, enabling richer data streams from sensors.
- **Low Latency:** Facilitates real-time responsiveness, critical for applications like autonomous vehicles and industrial automation.
- **Massive IoT Connectivity:** 5G can connect billions of devices simultaneously, making it ideal for AII's

scalability.

Security Measures in Wireless Data Transmission

As AII systems transmit sensitive data, robust security measures are essential:

- **Encryption:** Ensures that data is unreadable without proper decryption keys.
- **Authentication Protocols:** Validates the identity of devices and users to prevent unauthorized access.
- **Intrusion Detection Systems:** Monitors networks for suspicious activity and mitigates potential threats.

2.3 Edge and Cloud Computing Synergy

Edge Processing for Local Decision-Making

Edge computing involves processing data at or near the source of data collection, reducing reliance on centralized systems:

- **Faster Decisions:** Real-time analytics on edge devices minimize latency, critical for time-sensitive applications like health monitoring and disaster response.

- **Reduced Bandwidth Usage:** By processing data locally, edge devices send only essential information to the cloud, optimizing network efficiency.
- **Energy Efficiency:** Local processing reduces the energy demand associated with transmitting large volumes of data.

Cloud-Based Analytics for Large-Scale Insights

The cloud complements edge computing by offering powerful, centralized resources for analyzing vast datasets:

- **Scalability:** Cloud platforms can scale storage and computing resources dynamically to handle fluctuating workloads.
- **Comprehensive Insights:** Aggregating data from multiple sources enables pattern detection and predictive analytics across systems.
- **Integration with AI:** Cloud-based AI models enhance the depth of analysis, supporting applications like behavioral predictions and anomaly detection.

Balancing Latency, Bandwidth, and Computational Power

The interplay between edge and cloud computing must balance competing demands to optimize AII performance:

- **Latency-Sensitive Tasks:** Tasks requiring immediate action, such as environmental monitoring, are prioritized for edge processing.
- **Bandwidth Considerations:** Data-intensive applications, like video analytics, offload processing to the cloud to reduce strain on local devices.
- **Computational Allocation:** A hybrid approach ensures that both edge and cloud resources are used judiciously, maximizing efficiency and minimizing costs.

The integration of these core technologies enables AII to function as a seamless, efficient, and sustainable intelligence layer within our environments. By understanding the intricate workings of smart tags, wireless communication, and computing paradigms, we can appreciate the transformative potential of AII in reshaping industries and everyday life.

CHAPTER 3

Initial Uses in Retail

In the retail industry, ambient invisible intelligence (AII) is already demonstrating its revolutionary potential by transforming consumer satisfaction, inventory control, and operational effectiveness. The early uses of AII in retail are examined in this chapter, along with the effects and difficulties encountered throughout integration.

3.1 Stock and Inventory Control

Smart Sensor-Based Stock Checking Automation

Inventory management is the foundation of effective retail operations, and AII has ushered in a new era of automation:

- Stock levels are continuously monitored by sensors and RFID-enabled tags attached to products or shelves. When things are taken away, lost, or in need of restocking, these devices recognize it.

- **Removal of Manual Mistakes:** AII lowers human errors such as missing restocking or erroneous counts that are frequent in manual procedures by automating stock inspections.

Updates for Inventory Visibility in Real Time

AII systems guarantee dynamic, real-time inventory tracking:

- Retailers may monitor inventory data from all locations on a single platform thanks to centralized dashboards, which facilitate quicker decision-making and replenishment planning.
- **Predictive Analytics:** AII helps forecast demand by examining real-time data and past sales trends, preventing both overstocking and understocking.
- **Integration with Supply Chain:** Just-in-time deliveries are guaranteed by seamless connectivity with suppliers, which simplifies logistics and lowers holding costs.

Advantages for Large-Scale Retail Operations

- **Scalability**: AII's affordable sensors and smart tags are particularly helpful for big shops who are in

charge of huge inventory.
- Advanced tracking aids in preventing shrinkage brought on by administrative mistakes or theft.
- **Increased Efficiency**: Instead of performing manual stock inspections, automated solutions allow staff to concentrate on strategic work and customer service.

3.2 Improving the Client Experience

Customizing Using Data-Driven Insights

AII helps merchants provide incredibly customized purchasing experiences, which increases client happiness and loyalty:
- **Behavioral Analytics:** Sensors and smart tags monitor consumer behavior and preferences in the store, giving information about areas and popular items.
- **Personalized Suggestions:** Real-time, personalized product recommendations are made possible by data gathered from past purchases and in-store activities.
- **Interactive Displays:** Smart shelves with screens can make product recommendations based on a customer's preferences or profile, resulting in a

customized in-store experience.

Instances of Smart Shelves and Self-Checkout Systems

- **Smart Shelves:** These shelves use RFID technology and weight sensors to track inventory levels and provide customers pertinent product information.
- A smooth, cashierless checkout process is made possible by integrated AII, which enables users to scan things automatically as they are added to a shopping basket.
- **Contactless Payment:** Retailers offer quicker and more hygienic transactions by integrating AII systems with NFC-enabled payment options.

Difficulties in Integrating Ambient Intelligence with Legacy Systems

- **Compatibility Issues:** Since many merchants continue to use outdated point-of-sale systems, integrating AII technology can be difficult and expensive.
- **Data Silos:** It takes a lot of work to unify and synchronize data streams with ALL platforms because legacy systems frequently function in

isolation.
- **Employee Training:** Adopting AII requires staff training, which big operations may find resource-intensive.
- **Cost of Transition:** For small and medium-sized merchants in particular, upfront expenditures for new technology and infrastructure can be costly.

3.3 Cutting Down on Operating Expenses

Savings via Better Stock Control and Less Waste
AII-driven inventory systems optimize stock management and reduce waste, which have a direct effect on a retailer's bottom line:
- **Decreased Overstock:** Real-time inventory tracking and precise demand forecasting assist prevent excess stock that could otherwise result in markdowns or waste.
- **Perishable Goods Management:** Sensors keep an eye on humidity and temperature for grocery and food stores, guaranteeing ideal storage conditions and minimizing spoiling.
- **Streamlined Logistics:** AII lowers storage costs and

improves delivery efficiency by coordinating inventory with supply chain processes.

Case Studies Showing ROI in Retail Applications

- **Grocery Retailer:** To keep an eye on perishable goods, a major supermarket chain implemented AII systems. Within a year, the outcome was a 15% increase in profitability and a 20% decrease in food waste.
- **Clothing Retailer:** A fashion retailer reported a 10% increase in sales due to improved stock availability and a 25% decrease in misplaced inventory by installing smart fitting rooms and inventory tracking.
- **Warehouse Automation:** By integrating AII-driven robots and sensors, a warehouse was able to achieve a 40% faster order fulfillment rate while reducing labor costs by 30%.

In addition to lowering operating expenses, the use of ambient invisible intelligence in retail has improved customer interaction and brought previously unheard-of efficiencies. The long-term advantages—such as enhanced inventory control, customized shopping, and substantial cost savings establish AII as a disruptive force in the retail sector, despite ongoing hurdles like integration with older systems.

CHAPTER 4

Supply Chain Transformation and Logistics

Supply chain and logistics processes are being revolutionized by Ambient Invisible Intelligence (AII), which provides automatic efficiencies, improved transparency, and real-time monitoring. This chapter examines how AII simplifies warehouse management, guarantees transparency, and transforms tracking perishable commodities.

4.1 Monitoring Perishable Items

One essential component of contemporary supply chain management is the effective handling of perishable items. AII addresses issues related to spoiling and regulatory compliance by utilizing smart sensors and sophisticated analytics.

Tracking Temperature, Humidity, and Shelf-Life

Metrics

AII uses state-of-the-art sensors to track environmental factors that are essential for maintaining perishable goods:

- In order to provide ideal conditions for delicate items like food, medications, and flowers, sensors are incorporated into packaging or transportation containers to assess temperature, humidity, and light exposure.
- Dynamic Shelf-Life Tracking: This feature prioritizes the movement of products that are about to expire by using real-time data to compute remaining shelf-life.

Using Real-Time Alerts to Prevent Spoilage

When environmental conditions depart from permitted parameters, AII systems instantly notify users:

- **Cloud-Connected Alerts:** Logistics teams receive alerts through cloud platforms, allowing for quick actions to return situations to normal.
- **Predictive Maintenance:** AII can reduce the danger of spoiling by anticipating possible equipment failures in refrigeration units by evaluating past data.
- **Automated Adjustments:** To maintain optimal

conditions, temperature controls in sophisticated systems can make adjustments automatically based on sensor data.

Adherence to Safety and Regulatory Standards
Strict guidelines for managing perishable items are enforced by governments and trade associations. AII makes compliance easier by offering precise, auditable data:

- **Digital Records:** Automated data logging helps with regulatory inspections and certifications by guaranteeing comprehensive records of storage and transportation circumstances.
- **International Standardization Across Borders:** AII makes it easy to adhere to international standards, guaranteeing that products satisfy quality requirements in worldwide marketplaces.

4.2 Facilitating Traceability and Transparency

Transparency and traceability are essential for cutting down on inefficiencies, stopping fraud, and fostering stakeholder trust in a global supply chain.

Monitoring Products in International Supply Chains

AII makes it possible to track products from point of origin to point of destination:

- **Geolocation Sensors:** GPS-enabled smart tags offer real-time location information, cutting down on delays and enhancing accountability.
- The ability to log data from AII sensors into decentralized ledgers ensures an unchangeable and transparent record of movements and transactions. This is known as blockchain integration.

Using Invisible Markers to Stop Counterfeit Goods

From luxury products to pharmaceuticals, businesses have been hampered by the spread of counterfeit goods. AII successfully addresses this problem:

- **Advanced Authentication:** Specialized scanners can be used to confirm the authenticity of invisible identifiers implanted in goods or packaging.
- **Tamper-Detection Sensors:** Smart tags are able to identify and notify relevant parties in the event that a shipment has been tampered with while in transit.
- **Building Stakeholder Trust Through Improved**

Traceability: Increased transparency cultivates confidence among partners, consumers, and authorities:

- **Consumer Confidence:** Using NFC tags or QR codes, customers may obtain comprehensive product histories that confirm the legitimacy and place of origin of products.
- **Collaborative Insights:** AII-powered transparent data-sharing platforms encourage cooperation among distributors, manufacturers, and suppliers, minimizing inefficiencies and disagreements.

4.3 Simplifying the Management of Warehouses

The foundation of effective supply networks is warehouse operations. AII uses modern robotics, automation, and space optimization to enhance warehouse management.

Optimizing Space and Resource Allocation

AII facilitates the best possible use of resources and warehouse space:

- **Dynamic Inventory Mapping:** AI algorithms and sensors continuously evaluate inventory levels and recommend the best storage arrangements.

- **Load Balancing:** AII systems avoid overloading and underuse of storage spaces by distributing loads equally based on item weights and dimensions.

Automated Inventory Updates and Sorting

Automation increases accuracy and decreases manual labor in warehouse operations:

- **Smart Conveyors:** AII-enabled tags allow items to be automatically sorted by kind, destination, or priority.
- In order to reduce errors, inventory systems are updated in real-time as things enter and exit the warehouse.

Autonomous Vehicles and Robotics Integration

AII and robots work together to increase productivity and lower operating expenses.

- **Autonomous Guided Vehicles (AGVs):** Robots with AI-enabled sensors move items to predetermined spots in warehouses on their own.
- **Cobots, or collaborative robots, are:** These robots increase overall efficiency by assisting humans with

monotonous jobs like sorting, packing, and picking.

- **Drone Integration**: Drones with cameras and sensors can swiftly search inventory and find lost products in big warehouses.

Ambient Invisible Intelligence improves transparency, boosts efficiency, and guarantees adherence to international standards by revolutionizing supply chains and logistics. These developments enable companies to meet the demands of contemporary commerce while preserving a competitive edge by delivering items more quickly, safely, and sustainably.

CHAPTER 5

AII's Revolution in Healthcare

Ambient Invisible Intelligence (AII) is driving a major revolution in the healthcare sector. AII is transforming hospital administration, patient care, and assistance for disadvantaged groups by combining cutting-edge sensors, intelligent technologies, and real-time data processing. This chapter explores the many facets of AII's influence on healthcare, including resource optimization, patient monitoring, and support for the old and disabled.

5.1 Tracking Medication and Patient Monitoring

By enabling smart medication systems and continuous monitoring, AII provides innovative solutions for patient care, tackling some of the most enduring problems in healthcare.

Ongoing Health Tracking With Wearable Tags

AII-powered wearable sensors deliver real-time vital sign data, providing previously unheard-of insights into patient health:

- **Comprehensive Data Collection:** Heart rate, blood pressure, glucose levels, oxygen saturation, and other parameters are monitored via devices.
- **Seamless Integration:** AII-enabled wearables are lightweight and discrete, so patients can continue with their everyday activities uninterrupted. This is in contrast to large monitors.
- **Proactive Interventions:** Health care professionals can take action before problems become emergencies by responding to alerts set off by abnormal readings.
- **Rural Monitoring:** AII improves telemedicine by sending doctors real-time health data, allowing patients in underserved or rural places to receive prompt consultations.

Intelligent Drug Administration Systems to Lower Human Errors

Medication errors, which frequently arise from improper dosages or missed schedules, continue to be a major problem in the healthcare industry. AII uses intelligent systems to reduce these risks:

- **Automated Dispensing Units:** By utilizing AII sensors, these devices minimize the need for manual processes by delivering the appropriate medication at the appropriate moment.
- The Connected Pill Bottles are smart bottles that notify caretakers if a patient forgets to take their prescription.
- **Personalized Dosage Adjustments**: AII uses patient data analysis to suggest the best dosages, adjusting therapies to meet the needs of each patient.

Improving Post-Operative Care and Chronic Condition Management

Consistent monitoring and adherence to care plans are necessary for both postoperative recovery and chronic disease management. AII helps achieve these objectives by:

- **Surgical Site Monitoring:** Sensors monitor the healing process and identify infection early.
- **Chronic Condition Tracking:** AII offers real-time information for ailments including diabetes and

hypertension, assisting physicians and patients in better managing their conditions.
- **Integrated Health Dashboards:** Patients and medical professionals may view recovery metrics, medication schedules, and health trends on a single platform.

5.2 Optimization of Hospital Resources

In the healthcare industry, where delays or shortages can have fatal outcomes, effective resource management is essential. AII ensures better care delivery by streamlining hospital operations.

Tracking Assets for Vital Equipment
During emergencies, hospitals frequently struggle to find necessary equipment. AII uses sophisticated tracking to address this:
- **Real-Time Locators:** Real-time location data is provided by sensors that are connected to medical equipment like defibrillators and ventilators.
- **Utilization Analytics:** AII systems examine patterns of equipment use to find underutilized resources and

maximize distribution.

- Sensors track the performance of equipment, anticipating malfunctions and planning maintenance before problems occur. This is known as preventive maintenance.

Improving Operational Efficiency to Cut Down on Patient Wait Times

Long wait times have a detrimental effect on patient outcomes and satisfaction. AII improves productivity by simplifying processes:

- **Dynamic Scheduling:** Appointment scheduling is optimized by real-time data on staff availability and patient flow.
- **Coordination of the Emergency Department:** AII systems rank patients according to urgency, guaranteeing that urgent cases are attended to right away.
- **Smart Bed Management:** Hospitals can more efficiently allocate resources by using sensors to determine bed availability.

Real-World Healthcare Deployment Examples

AII is being used by several healthcare organizations to enhance results:

- **Johns Hopkins Hospital:** Reduces operational bottlenecks by using AII for real-time equipment tracking and patient monitoring.
- **Singapore General Hospital:** Uses AII-enabled devices to reduce surgery wait times and maximize operating room utilization.
- **Innovations from Startups:** Wearable health monitors with ALL platforms are being pioneered by companies such as Medtronic and Biofourmis.

5.3 Assisting the Elderly and Disabled

For the aged and disabled, AII is revolutionary because it promotes independence while guaranteeing safety through intelligent monitoring systems.

Detection of Fall and Emergency Notifications

One of the main causes of injuries among the elderly is falls. AII reduces these dangers by:

- **Wearable Accelerometers:** These sensors automatically notify emergency services or

caregivers when they detect abrupt movements that could be indicative of a fall.
- Floor pressure sensors are used by AII-enabled smart houses to identify falls and sound alarms.
- **Customized Response Protocols:** Depending on the incident's seriousness, alerts can be customized to alert emergency personnel, family members, or medical professionals.

Smart Home Technologies for Increased Independence

AII-powered smart homes allow the elderly and disabled to live independently while maintaining contact with caregivers:

- **Voice-Activated Controls:** AII-integrated systems enable voice commands to operate appliances, lighting, and security functions.
- **Health Monitoring:** Sensors built into chairs and beds monitor posture, activity levels, and sleep quality to provide important health information.
- **Automated Assistance:** Safety and convenience are improved by features like automated door locks, fall prevention technologies, and prescription reminders.

Ethical Aspects of Pervasive Monitoring Deployment

Even though AII has many advantages, there are some moral concerns that need to be addressed:

- **Privacy Concerns:** Constant observation can seem invasive, thus precise rules about data gathering and use are required.
- **Consent and Autonomy:** To ensure that users keep control over their life, they should be able to choose whether to use certain features or not.
- Developers must make sure ALL systems are inclusive and free of biases that can disfavor particular communities.
- **Balancing Safety and Freedom:** Ethical AII deployment requires striking the correct balance between personal freedom and safety monitoring.

By boosting the quality of life for vulnerable groups, optimizing hospital resources, and increasing patient outcomes, ambient invisible intelligence is transforming the healthcare industry. To guarantee that emerging technologies provide advantages in a fair and responsible manner, stakeholders must manage ethical issues as use rises. AII has the ability to completely transform healthcare and usher in a new era of patient-centered, intelligent care.

CHAPTER 6

Conservation and Monitoring of the Environment

Ambient Invisible Intelligence (AII) is revolutionizing environmental monitoring and protection. AII improves pollution monitoring, disaster management, and animal preservation by fusing smart sensors and real-time data analytics with environmental conservation. This chapter examines how AII is promoting sustainable practices and environmental stewardship.

6.1 Monitoring and Preserving Wildlife

One urgent worldwide concern is the loss of biodiversity. AII provides cutting-edge resources for monitoring, safeguarding, and conserving endangered animals and their environments.

Tracking Endangered Species using AII

Tracking tools and sensors with AII capabilities offer

comprehensive insights into animal behavior and migration patterns:

- **GPS and Sensor Integration:** Animals are equipped with lightweight tags that monitor their movements, migration paths, and preferred habitats.
- **Behavioral Analysis:** In order to identify possible dangers like poaching or environmental disturbances, intelligent algorithms examine patterns of activity.
- **Remote Monitoring:** By monitoring species remotely, conservationists can lessen their reliance on intrusive on-ground tracking techniques.

Data-Driven Approaches to Habitat Protection

AII makes it possible to gather accurate data that guides habitat conservation plans:

- **Identifying Critical Habitats:** Sensors pinpoint locations that are necessary for refuge, food, or breeding, enabling focused preservation efforts.
- In order to avert conflicts and protect both communities and wildlife, early warning systems identify when animals are approaching human settlements.

- **Long-Term Ecological Monitoring:** AII offers ongoing data that aids scientists in comprehending long-term environmental shifts and how they affect species.

Complementing Conservation Initiatives Worldwide

The influence of AII on wildlife conservation is increased when governments, non-governmental organizations, and private groups work together:

- **Global Data Sharing:** AII data is integrated by platforms such as the Global Biodiversity Information Facility (GBIF), facilitating global conservation initiatives.
- **Community Engagement:** By empowering local communities to employ AII tools, grassroots conservation efforts are fostered.
- **Success Stories:** Initiatives such as tracking the movements of African elephants by collaring them and monitoring poaching hotspots with drones fitted with ALL sensors show how revolutionary AII can be.

6.2 Monitoring of Air and Water Quality

Both human health and ecosystems are seriously threatened by pollution. AII provides reliable real-time monitoring and mitigation solutions.

Collection and Visualization of Real-Time Pollution Data

AII offers detailed, up-to-date information on the quality of the air and water, facilitating prompt solutions to environmental issues:

- **Air Quality Sensors:** These devices measure pollutants at the neighborhood level, such as carbon monoxide (CO), nitrogen dioxide (NO_2), and particulate matter (PM2.5).
- **Water Quality Monitors:** Sensors identify pollutants in water bodies, including microbial loads, heavy metals, and pH abnormalities.
- The usage of AII-powered platforms to present pollution data in easily comprehensible ways allows citizens and policymakers to comprehend and take action on environmental problems.

Urban Planning and Public Health Applications

AII-driven insights inform health interventions and sustainable urban development:

- **Smart Cities:** Industrial zoning, traffic control systems, and green space placement are all influenced by air quality data.
- **Public Health Alerts:** Health alerts are triggered by real-time pollution monitoring, which lowers exposure risks during pollution surges.
- **Identification of Pollution origins:** AII identifies the origins of contamination, enabling focused enforcement and cleanup efforts.

The Function of the Public and Private Sectors in Using AII

For ALL applications to scale, cooperation between public and private institutions is essential:

- **Policy Development:** Lawmakers have the authority to enact rules for the application of AII in urban and industrial surveillance.
- **Corporate Responsibility:** Businesses can lessen their environmental impact by incorporating AII into sustainability activities.

- **Funding and Innovation:** Public-private partnerships and investments in AII startups hasten the creation of innovative monitoring solutions.

6.3 Early Warning Systems and Disaster Management

Natural disasters result in enormous damage and fatalities. By offering precise, up-to-date information, AII improves catastrophe preparedness, response, and recovery.

Using Sensor Networks to Predict Natural Disasters

AII uses vast networks of sensors to identify and forecast natural hazards:

- **Seismic Activity Monitoring:** Sensors identify subtle tectonic movements, providing earthquake warnings in advance.
- **Weather Pattern Analysis:** AII forecasts storms, floods, and droughts by analyzing data from weather sensors.
- **Environmental Indicators:** To predict disasters, changes in soil moisture, river levels, and meteorological conditions are regularly tracked.

Improving Situational Awareness to Coordinate Relief Efforts

AII facilitates accurate, real-time coordination, which enhances disaster response:

- **Resource Allocation:** Personnel, food, and medical supplies are deployed to regions of highest need based on real-time data.
- Even in places where infrastructure is compromised, first responders can communicate reliably thanks to AII-driven platforms.
- The integration of AII insights with citizen-generated data improves situational awareness and response planning.

AII Success Stories in Disaster-Prone Areas

AII has shown impressive results in reducing the effects of disasters:

- **Tsunami Early Warning Systems:** AII networks are used in coastal areas of Japan and Indonesia to deliver timely tsunami warnings, which have prevented thousands of fatalities.
- **Flood Monitoring in India:** River sensors with AII capabilities assist authorities in anticipating and

controlling floods, minimizing fatalities and financial damages.

- **Hurricane Management in the United States:** AII systems monitor the trajectories and intensities of storms, directing emergency preparations and evacuation strategies.

Through the incorporation of AII into environmental monitoring and conservation initiatives, communities may effectively and strategically tackle urgent ecological issues. AII paves the way for a more sustainable future by enabling stakeholders to defend biodiversity, fight pollution, and improve catastrophe resilience.

CHAPTER 7

Smart Cities and Urban Infrastructure

The future of urban living is being redefined by Ambient Invisible Intelligence (AII), which makes cities safer, more effective, and more sustainable. Cities are becoming dynamic ecosystems that adjust to the requirements of their residents by incorporating AII into their transportation, energy management, and public safety systems. This chapter explores how AII is revolutionizing smart cities and urban infrastructure.

7.1 Improving Safety for the Public

A key component of smart city development is public safety, and ALL is essential to creating safe and resilient urban settings.

Incorporation of AII into Monitoring Systems

Surveillance systems driven by AII improve situational

awareness and quick reaction times:

- **Smart Cameras:** Outfitted with sophisticated analytics, these cameras identify anomalous activity, like abnormal conduct or unattended baggage.
- **Facial Recognition:** Although controversial, this technology can identify people engaged in crimes or missing persons situations when used appropriately.
- **Distributed Sensor Networks:** To improve community safety, sensors integrated into public infrastructure keep an eye on environmental factors like air quality and dangerous gas leaks.

Keeping an Eye on Crowd Density to Avoid Mishaps

In metropolitan areas, crowd control is crucial, particularly during emergencies or events:

- **Density Tracking:** AII sensors use real-time crowd density analysis to pinpoint places that could experience traffic jams or stampedes.
- **Evacuation Planning:** During emergencies, integrated systems recommend the best evacuation routes, which lessens confusion and fear.
- **Event Management:** AII helps planners manage foot traffic and reduce hazards at big events.

Regulatory Frameworks and Privacy Concerns

There are important privacy concerns raised by the extensive usage of AII in public safety:

- **Data Anonymization:** Algorithms must guarantee the anonymization and security of personal data gathered by surveillance systems.
- **Legislation and Oversight:** To avoid abuse, governments and regulatory agencies must set precise rules for the use of AII in public safety.
- **Public Engagement:** Including the public in conversations about surveillance technology promotes openness and confidence.

7.2 Energy and Utilities Optimization

Two essential elements of sustainable urban development are resource management and energy efficiency. AII enables creative approaches to resource optimization and utility management.

Smart Grid Control AII converts energy distribution networks into networks that are responsive and effective:

- **Demand Response**: By adjusting the energy supply in real-time in response to demand, smart meters and sensors minimize waste and prevent blackouts.
- **Renewable Integration:** AII maximizes the grid's ability to integrate renewable energy sources like wind and solar.
- Sensors identify possible problems in energy infrastructure, allowing for preventive repairs and minimizing downtime. This is known as predictive maintenance.

Innovations in Water and Waste Management

Urban sustainability depends on effective waste and water management:
- **Leak Detection:** AII-powered sensors keep an eye out for leaks in water pipelines, reducing maintenance costs and limiting water loss.
- **trash Sorting:** Reducing landfill contributions, recyclables are sorted using smart trash bins with sensors and AII algorithms.
- **Consumption Monitoring:** Businesses and citizens are encouraged to adopt sustainable practices by real-time data on water usage.

Energy-Efficient City Case Studies

AII is being used by cities worldwide to increase energy efficiency:

- **Amsterdam:** The city uses AII for smart street lighting, which uses less energy by dynamically modifying light levels according to traffic, both pedestrian and vehicular.
- Despite having few natural resources, Singapore's sophisticated AII systems provide a steady supply of water.
- **San Diego:** AII-powered microgrids help the city lessen its need on non-renewable energy sources and maintain energy resiliency.

7.3 Intelligent Transportation and Traffic Management

The foundation of smart cities is an effective transportation infrastructure. AII makes it easier to integrate autonomous vehicles, improves traffic management, and maximizes public transportation.

Real-Time Traffic Flow Analysis AII uses intelligent

monitoring technologies to assist authorities and urban planners in reducing traffic congestion:

- **Traffic Lights**: These sensors, which are placed along roadways and at intersections, provide information on vehicle counts, speeds, and areas of high traffic.
- **Adaptive Traffic Lights:** AII-enabled signals modify timings in response to traffic patterns, enhancing flow and cutting down on delays.
- **Accident Prediction:** Preventive actions are made possible by real-time analysis that pinpoints high-risk locations and circumstances.

Applications in Public Transit Optimization

AII improves the dependability and usability of public transportation systems:

- **Dynamic Scheduling:** Train and bus timetables adjust to current demand, reducing crowding and wait times.
- **Predictive Maintenance:** Sensors keep an eye on infrastructure and transit vehicles to avoid malfunctions and interruptions in service.
- **Passenger Experience:** AII apps offer up-to-date

information on delays, available seats, and other routes.

Autonomous Cars and AII Partnership

Urban mobility is being revolutionized by the partnership of AII and autonomous car technologies:

- **Traffic Ecosystem Integration:** AII data is used by autonomous cars to safely negotiate intricate traffic situations.
- **Shared Mobility Services:** AII helps ride-sharing services assign cars to users in real time, cutting down on empty miles and wait times.
- **Safety Enhancements**: Sensors identify cyclists, pedestrians, and other vehicles, guaranteeing safe passage through crowded urban areas.

Cities can make notable improvements in mobility, resource efficiency, and public safety by integrating AII into their urban infrastructure. These developments create the foundation for more resilient and sustainable communities while also enhancing the enjoyment of urban life.

CHAPTER 8

Everyday Life and Consumer Technology

Customers' interactions with technology are being revolutionized by Ambient Invisible Intelligence (AII), which blends in seamlessly with daily activities to improve productivity, comfort, and convenience. AII-driven solutions are changing modern living in a variety of ways, including smart homes, tailored user experiences, and adaptive learning environments.

8.1 Wearable technology and smart homes

AII is at the heart of the wearable technology boom and the smart home revolution, both of which significantly improve daily life.

Home Appliances and Utilities Automation
AII is used by smart home systems to automate and optimize domestic tasks, decreasing manual labor and

increasing energy efficiency:

- **Energy Management:** Nest and Ecobee smart thermostats learn user preferences and make dynamic temperature adjustments to save energy and maintain comfort.
- **Lighting Systems:** AII algorithms and motion sensors drive automated lighting that modifies brightness according to occupancy and daylight.
- **Kitchen Automation:** When supplies are running low, smart refrigerators automatically reorder necessities, keep an eye on inventory, and recommend meals depending on the items that are available.

Improving User Experiences with Predictive Behavior Models

AII systems create a customized experience by learning and adapting to user behavior:

- **Daily Routines:** Automated devices anticipate user demands, like locking doors at night or turning on the coffee maker in the morning.
- **Voice Assistants:** AII is used by systems such as Google Assistant and Alexa to improve their

responses and recommend pertinent actions based on past usage trends.

- **Health Monitoring**: AII is used by wearable technology, such as the Apple Watch and Fitbit, to track fitness data and offer tailored health advice.

Difficulties with Interoperability in Smart Home Environments

Despite their advantages, smart homes have trouble integrating all of its equipment seamlessly:

- **Diverse Protocols:** Compatibility problems frequently arise when Bluetooth, Zigbee, and Wi-Fi technologies coexist.
- **Vendor Lock-In:** A lot of manufacturers create systems that are exclusive to their ecosystems, which restricts the options available to customers.
- **Data Synchronization:** It is still technically difficult to guarantee seamless communication across devices without interruptions or conflicts.

8.2 Tailored User Interfaces

AII is enabling previously unheard-of degrees of

customisation in consumer electronics, improving privacy, entertainment, and general wellbeing.

Adaptive AII Applications in Entertainment Examples

AII is used by entertainment platforms to provide highly customized content:

- **Streaming Services:** AII is used by platforms such as Netflix and Spotify to examine viewing or listening patterns and provide personalized recommendations.

Gaming

- Adaptive AI in games adjusts storylines and difficulty levels to player performance and preferences.
- **Virtual Reality (VR):** AII improves VR experiences by dynamically modifying virtual worlds to fit user behavior and emotions.

Convenience and user privacy are balanced.

A cautious approach to privacy is necessary due to the growing personalization of AII systems:

- Strong encryption procedures guarantee that private user information is protected against intrusions.

- **Transparency:** Businesses must explain in detail how they gather, handle, and use user data.
- **User Control:** It's crucial to provide users authority over their data and the choice to refuse certain services.

Participation in Encouraging Mental and Physical Health

Devices with AII capabilities are significantly enhancing wellbeing:

- **Mental Health Apps:** Apps such as Calm or Headspace promote mindfulness by tailoring their programs to the stress levels and preferences of their users.
- **Sleep Trackers:** Wearable technology tracks sleep habits and recommends lifestyle modifications for improved sleep.
- **Health Interventions:** By warning users of abnormal heart rates, inactivity, or other health issues, smart gadgets allow for prompt medical intervention.

8.3 AII in Learning and Education

Another area where AII is generating transformative opportunities and promoting more effective and engaging learning experiences is education.

Adaptive Learning Experiences with Smart Classrooms
AII-powered classrooms accommodate a range of learning requirements and preferences:

- **Personalized Lessons:** Programs evaluate each student's development and modify the material to fit their comprehension level and speed.
- **Interactive Learning:** AII-enabled educational software and interactive whiteboards are examples of technologies that enhance learning.
- **Remote Accessibility:** By providing immersive and interactive experiences, smart tools help close the gap for learners who are located far away.

Student involvement Assessment in Real Time
AII improves the capacity to track and raise student involvement in real time:

- **Facial Recognition:** To determine kids who are not

paying attention, sensors examine facial expressions.

- **Feedback Mechanisms:** Teachers get useful information that enables them to modify their teaching strategies at any time.
- Exams that use adaptive testing dynamically change in difficulty according to student performance, giving a more realistic picture of a student's aptitudes.

Setting Up Organizations for a Smooth AII Integration

Institutions need to meet a number of conditions for AII to flourish in educational settings:

- **Infrastructure Upgrades:** To enable AII systems, schools require smart devices and strong internet access.
- To effectively use ALL tools and comprehend their potential, educators need to receive training.
- **Ethical Principles:** All students will profit from AII if standards for data privacy and fair access are established.

AII has a significant impact on consumer technology, improving daily living in terms of efficiency, engagement, and health consciousness. AII may continue to influence a future in which technology is smoothly incorporated into our personal and professional life by tackling the issues of accessibility, privacy, and interoperability.

CHAPTER 9

Adoption of AII: Obstacles and Hazards

There are many chances for creativity, efficiency, and convenience as Ambient Invisible Intelligence (AII) develops and spreads throughout different industries. To ensure its responsible and sustainable expansion, a number of risks and obstacles associated with the adoption and integration of AII must be carefully considered. These difficulties include problems with technology infrastructure, privacy, and morality. This chapter delves further into these issues, highlighting the challenges that need to be overcome in order to realize AII's full potential.

9.1 Security and Privacy Issues

Continuous data collecting through sensors and linked devices is a necessary part of integrating AII into daily life, which invariably brings up serious privacy and security concerns. Understanding the possible hazards connected to the enormous volumes of data being produced is essential as AII systems proliferate.

Possible Abuse of Widespread Tracking Information

AII systems work by continuously monitoring people, things, and environments often in real time. If not handled appropriately, this massive data collection could have unexpected consequences:

- **Data Exploitation:** Businesses and outside parties may abuse private information for profit, such as by influencing customer behavior or using personal information to target ads without permission.
- **Surveillance Overreach:** People's feeling of privacy and autonomy may be compromised by situations where they feel like they are being watched all the time due to widespread tracking. This is especially problematic in semi-private or public settings because AII systems gather personal information about individuals without their knowledge.
- The centralized collecting and storage of sensitive data increases the risk of data breaches, which might give malevolent actors access to location-based and personal information.

Creating Sturdy Anonymization and Encryption Methods

Strong data protection methods must be incorporated into the system architecture in order to reduce the privacy threats associated with AII:

- **End-to-End Encryption:** Protecting sensitive data from unwanted access and ensuring that it is encrypted both during transmission and storage.
- **Anonymization:** While preserving the ability to use data for more general analytical purposes, data anonymization techniques, like eliminating personally identifiable information (PII), can lower the risk of privacy violations.
- **Secure Authentication:** By putting robust multi-factor authentication systems in place, sensitive data is protected from unauthorized access by authorized individuals.

Regulatory Actions to Guarantee Appropriate Use

To control AII use and guarantee the protection of privacy rights, governments and regulatory agencies must set up frameworks:

- **Data Protection Regulations**: Enforcing regulations such as the General Data Protection Regulation (GDPR) in the EU can give consumers legal protections and guarantee that businesses handle personal data in an appropriate manner.
- **Transparency and Accountability**: To enable consumers to make knowledgeable decisions regarding their privacy, regulations should mandate that businesses reveal the ways in which data is gathered, utilized, and shared.
- Independent audits of AII systems can assist guarantee that privacy requirements are being respected and that any possible misuse is quickly detected and fixed.

9.2 Barriers Caused by Technology

Despite AII's enormous potential, a number of technological obstacles prevent its broad use and efficient implementation. In order for AII to realize its full potential, certain obstacles pertaining to sensor networks, energy efficiency, and system compatibility must be removed.

Difficulties with Sensor Network Scaling

The infrastructure for AII applications requires a lot of resources and is technically challenging to build and maintain:

- The implementation and upkeep of extensive sensor networks in diverse settings, like cities or industrial sites, entail substantial expenses for both initial setup and continuing maintenance.

- **Management of Data:** It gets more difficult to manage and process the growing amount of data produced by sensor networks. To guarantee that sensor inputs provide useful insights, effective data storage, real-time processing, and analytics capabilities are crucial.

- **Network Reliability:** Environmental factors like interference, signal deterioration, or network congestion can impact sensor network performance, especially in remote or strongly populated areas. For AII systems to operate correctly, it is essential that these networks continue to be strong and dependable over time.

Trade-offs between Battery Life and Energy Efficiency

A lot of AII systems depend on energy-consuming sensors and devices, thus battery life and system performance are frequently traded off:

- **Power Consumption:** Wearable technology and remote sensors have a limited operating time due to the high energy consumption of sophisticated sensors and processing equipment.
- **Battery Technology:** Despite advancements in battery technology, longer-lasting, more effective batteries are still required to power ALL devices without the need for frequent replacements or recharging.
- In order to overcome these constraints, AII systems can include energy-harvesting technologies, including thermal, kinetic, or solar energy collecting, to prolong the lifespan of sensors and devices without exclusively depending on conventional batteries.

Differences in Compatibility Among Systems

Because ALL depends on a variety of technologies, maintaining compatibility across different devices and platforms continues to be a major challenge:

- The integration of AII technologies into current infrastructures may be hampered by devices from different manufacturers using incompatible data formats or communication protocols.
- **Standardization:** It is challenging to guarantee that AII systems from various suppliers function together without hiccups because of the absence of generally accepted standards for sensor connectivity and data exchange. To promote broad adoption, industry-wide initiatives to standardize technologies will be essential.
- **Legacy Systems:** Many sectors continue to use outdated infrastructure that could be difficult to integrate with AII technologies. It could be expensive and time-consuming to update or replace these systems in order to interact with more recent technologies.

9.3 Implications for Ethics and Society

The use of AII raises a number of ethical and societal issues, much like any other disruptive technology. In order to guarantee that AII serves society in a just and equitable way, it is imperative that these consequences be addressed.

The dangers of relying too much on automated systems

The widespread usage of AII may result in an excessive dependence on automated systems, which would reduce human judgment and agency:

- **Loss of Human Oversight:** Although AII systems are intended to make judgments on their own using real-time data, an excessive dependence on them may compromise human oversight and decision-making in crucial circumstances.
- **Dehumanization:** Excessive automation can result in a lack of personal touch in industries like healthcare or customer service, which lowers the standard of human interactions.
- **Skill Degradation:** People may lose important skills when automated systems take on more work, which reduces human competence and makes people more reliant on machines.

Resolving Biases in the Gathering and Evaluation of Data

Because AII systems rely so largely on data, they may

produce erroneous results if the data is biased:

- **Bias in Data Sources:** Social biases like gender, race, or socioeconomic status may be reflected in the data used to train AII models. These biases have the potential to sustain prejudice in decision-making processes if they are not addressed, especially in delicate domains like law enforcement, lending, and employment.
- **Algorithm Transparency:** To detect and correct biases in decision-making, AII algorithms must be clear and comprehensible.
- The utilization of representative and diverse data sets is essential for training AII systems in order to prevent bias and guarantee that the technology functions fairly for all populations.

Providing Fair Access to AII Advantages

It is crucial to make sure that everyone has equal access to these advantages as AII systems become more integrated into daily life:

- **Digital Divide:** Inequalities in access to smart gadgets, high-quality internet, and other digital tools may result in unequal AII advantages. Existing disparities may be exacerbated if marginalized areas lack the resources to adopt or profit from modern technologies.

- Affordability: Some groups of people may find the cost of implementing AII solutions to be unaffordable, especially in low-income or developing nations. To guarantee broad adoption, efforts must be made to lower the price of AII technologies and make them accessible.
- A truly egalitarian technology landscape requires that AII systems be made inclusive and accessible to individuals with impairments, those speaking different languages, and those from a variety of cultural backgrounds. This is known as inclusive design.

Ambient invisible intelligence adoption offers both enormous potential and formidable obstacles. To guarantee that this revolutionary technology serves society in a responsible and sustainable way, it is imperative to address privacy concerns, remove technological obstacles, and take into account the ethical and social ramifications of AII implementation. Stakeholders can overcome these obstacles and fully utilize AII by concentrating on accountability, openness, and accessibility.

CHAPTER 10

Ambient Invisible Intelligence's Future Prospects

Ambient Invisible Intelligence (AII) is a cutting-edge technical advancement that holds out the prospect of a time when the digital and physical realms coexist together. It is anticipated that as AII develops further, its incorporation into daily life will quicken and result in significant shifts in both sectors and communities. The possible future of AII is examined in this chapter, with particular attention paid to the development of a sustainable AII ecosystem, the major trends that will emerge until 2027, and its deeper integration into everyday life.

10.1 Greater Incorporation into Daily Life

According to AII's vision, intelligence will not only be present in individual gadgets but will also permeate every aspect of our lives and influence how we engage with both technology and the natural world. AII has the ability to improve human capacities, produce previously unheard-of efficiency, and revolutionize daily experiences as this vision comes to fruition.

The Goal of an Intelligent and Completely Connected Ecosystem

AII powers the future's completely linked ecosystem, in which all systems, environments, and devices are seamlessly connected, sharing data continuously, and streamlining procedures. This would enable real-time modifications to suit individual demands and the constant flow of information.

- **Smart Cities and Homes:** Infrastructure, roads, and buildings will be networked together to anticipate demands using real-time data. Based on tenant behavior, smart houses will automatically modify the temperature, lighting, and even the way food is prepared. In a similar vein, without human assistance, urban infrastructure will adjust to traffic patterns, energy consumption, and public services.

- **Personalized Environments:** Picture walking into a space where AII systems have learnt your preferences over time to automatically modify the lighting, temperature, and even background noise. This will go beyond residences and places of employment to public areas like dining establishments, airports, and medical institutions, establishing dynamic settings that meet the unique requirements of each person.

- **Wearable Technology Integration:** The AII ecosystem will increasingly incorporate wearables, including smart glasses, health monitors, and augmented reality (AR) gadgets. In addition to keeping an eye on health, these gadgets will also predict medical issues before they arise, connect with other smart devices to modify the surroundings, and provide tailored wellbeing advice.

Ubiquitous Intelligence: Enhancing Human Potential

The ability of deeper AII integration to enhance human capacities is one of its main advantages. Artificial intelligence (AII) can enhance human decision-making, increase productivity, and lessen cognitive overload by supplying continuous, invisible intelligence.

- **Cognitive Augmentation:** AII systems will assist people in making decisions by providing timely, contextually relevant information. This might include giving consumers real-time, individualized buying suggestions or assisting physicians in making diagnoses based on real-time data analysis.

- **Optimizing Productivity:** Without conscious human effort, AII helps with everyday chores, plans timetables, and offers pertinent insights for complex projects, which could result in considerable productivity gains for professionals. AII has the potential to support human creativity in creative professions by assisting with idea generation, prototype design, and work refinement.

AII systems may develop emotional and social intelligence in the future, which would allow them to comprehend and react to human emotions. This would promote more sympathetic and encouraging interactions between people and technology. This has the potential to revolutionize sectors where emotional and individualized connections are critical, such as healthcare, customer service, and education.

Futuristic AII Application Examples

Looking ahead, a number of cutting-edge AII applications

offer an idea of what can be achievable in a globalized world:

- **Smart Healthcare Systems:** By continuously monitoring patient vitals, surroundings, and past data, AII could anticipate health emergencies before they occur. When necessary, these devices might notify medical personnel and automatically modify prescription or treatment schedules.

- **Environmental Adaptation:** Picture a world in which AII systems continuously monitor and modify environmental factors like the weather or pollution levels. Based on real-time environmental data, smart cities and buildings might modify energy use, ventilation, and even access to green areas, enhancing quality of life and cutting down on resource waste.

- **Immersive Experiences:** AII has the potential to develop completely immersive virtual worlds that are customized to each user's tastes in the entertainment industry. AII's ability to produce dynamic, customized experiences will influence the future of entertainment, from mood-based movie suggestions to user-driven virtual vacation experiences.

10.2 Patterns Through and After 2027

Advances in AI, machine learning, sensor technologies, and the growing need for smarter systems are all contributing to the fast changing AII scene. The following patterns are anticipated to shape AII's destiny by 2027 and beyond.

Important Developments in Sensor Technologies
The foundation of AII is sensors, which supply the information required to build intelligent systems. We shall witness advancements in sensor technology's capabilities and environmental integration as it develops further.

- **Miniaturization and Cost Reduction:** Sensors will continue to get smaller, cheaper, and more energy-efficient, which will lead to their widespread use in everything from large-scale infrastructure projects to consumer electronics.
- **Advanced Sensing Capabilities:** New sensor technologies, such bio-sensors and hyperspectral sensors, will offer more profound understandings of

ambient changes, human health, and even emotional states. AII systems will be able to collect more detailed information and make wiser decisions as a result.

- Real-time, extensive deployment of smart devices and sensors across industries will be made possible by the launch of 5G networks, which will supply the bandwidth needed for the huge data exchange needed by AII systems.

AI and Machine Learning's Expanding Role in AII

The operation of AII systems will continue to be mostly driven by machine learning (ML) and artificial intelligence (AI). With the help of these technologies, AII will be able to change, grow, and learn over time, creating new chances for personalization and improvement.

- **Autonomous Learning:** As AII systems advance, they will no longer rely solely on pre-programmed reactions; instead, they will become completely autonomous, learning from their interactions and surroundings to gradually enhance their performance.
- **Predictive Capabilities:** AII systems will be able to

more accurately forecast future requirements, environmental changes, and human behavior thanks to machine learning. This might transform sectors like healthcare, logistics, and transportation by predicting traffic patterns and health issues before they happen.

Economic Impacts Forecast by Industry

AII adoption will have significant economic ramifications for a number of industries. The economic environment will probably be shaped by the following trends:

- AII will automate a lot of tasks, but it will also lead to the creation of new sectors and job opportunities. As the need for these technologies increases, positions in data science, AI development, and sensor network administration will become increasingly important. But this change will also necessitate retraining and reskilling employees in automation-threatened industries.
- **Efficiency Gains**: AII adoption will result in efficiency gains for industries, which will lower costs and boost profitability. Better process optimization, predictive maintenance, and

customized services will help the manufacturing, logistics, and healthcare industries.
- **Market Expansion:** AII technologies will open new markets, especially in poorer nations, as they become more accessible and affordable. This could propel global economic development by generating economic growth in previously underserved places.

10.3 Establishing an Ecosystem of Sustainable AII

Building a sustainable ecosystem that strikes a balance between technological advancement and environmental, social, and economic factors must go hand in hand with the wider deployment of AII.

Guides for Eco-Friendly Implementations

Policies and tactics that reduce environmental effects and encourage resource conservation will be necessary for the deployment of AII in a sustainable manner.
- **Energy-Efficient Systems:** Energy efficiency will be a major concern as AII systems become more integrated into everyday life. AI-driven energy optimization systems, energy-harvesting

technologies, and low-energy sensors will all be essential in lowering AII's environmental impact.

- **Models of the Circular Economy:** With an emphasis on recycling, reusing, and cutting down on electronic waste, the lifecycle of ALL devices and sensors must be managed within the framework of the circular economy. Manufacturers will be urged to use recyclable and sustainable materials and to design products with lifetime in mind.
- **Eco-Friendly Manufacturing:** Regulations supporting environmentally friendly production methods ought to go hand in hand with the development of AII systems. To make sure that the growth of AII does not worsen the environment, it will be essential to source resources sustainably, cut production carbon emissions, and minimize waste.

Public and Private Stakeholder Collaboration

Governments, commercial businesses, and civil society organizations must work together to create a sustainable AII ecosystem.

- **Public-Private Partnerships:** To finance the construction of AII infrastructure, guarantee fair access, and enact laws that advance sustainability, safety, and privacy, governments and businesses must collaborate.

- **International Collaboration:** International collaboration will be crucial to addressing issues with standards, ethics, and data security when AII technologies are deployed globally. Global rules and agreements will guarantee that AII serves every region fairly while lowering the dangers of abuse or unequal access.

- **Inclusive Innovation**: To guarantee that AII systems meet a range of demands, communities must be included in their creation. AII solutions will be made more accessible and relevant through community-driven innovation, partnerships with academic institutions, and public consultations.

Encouraging Innovation While Resolving Social Issues

Addressing societal difficulties such as privacy hazards, digital divisions, and ethical dilemmas must coexist with AII innovation.

- The development of AII systems must adhere to ethical standards in order to avoid unforeseen repercussions like bias or discrimination in AI algorithms. Building trust in these technologies will be facilitated by open and transparent AI development procedures.

- **Bridging the Digital Divide:** It is important to make sure that ALL technologies are available to everyone, irrespective of socioeconomic background or geographic location. Promoting global inclusion will require equitable access to AII instruments.

- **Privacy and Data Security:** Since AII systems gather enormous volumes of personal data, it will be crucial to have strong data protection measures in place to uphold public confidence and guarantee moral technology use.

Ambient Invisible Intelligence has the potential to completely transform how we work, live, and engage with the world. We can make sure that AII improves our daily lives in a way that is ethical, sustainable, and inclusive for everyone by embracing these trends and challenges.

Glossary

The following glossary covers important terms and concepts associated with Ambient Invisible Intelligence (AII) in order to maintain consistency and clarity across this book:

1. **Ambient Invisible Intelligence (AII).**
   a smooth incorporation of artificial intelligence, ultra-low-power systems, and smart sensing technologies into daily settings. AII improves productivity and user experience by discreetly gathering, analyzing, and reacting to data in real time.

2. **Smart Tags** Small, reasonably priced gadgets with wireless connection and sensor integration. Numerous applications, such as inventory management and healthcare, use these tags for tracking, monitoring, and data collection.

3. **Sensors with Ultra-Low Power**
   sensors with low power requirements, frequently

employing energy-harvesting methods. For AII systems to be sustainable and scalable, certain sensors are essential.

4. Data transmission between devices in an AII network is made possible by standards and technology known as Wireless Communication Protocols. Near-Field Communication (NFC), Bluetooth Low Energy (BLE), and Radio Frequency Identification (RFID) are a few examples.

5. **Fifth, Edge Computing**

    a decentralized computing paradigm that minimizes latency and bandwidth consumption by processing data locally, at the point of generation (such as sensors or devices).

6. **Cloud Computing**

    a centralized computer paradigm in which distant servers handle, store, and analyze data. Cloud computing is utilized in AII to supplement edge computing with large-scale analytics and storage.

7. **Watching in Real Time**

   the ongoing monitoring and evaluation of data as it is gathered, which facilitates prompt decision-making and action in fields including environmental management, healthcare, and transportation.

8. **The eighth is the Internet of Things (IoT).**

   a system of linked sensors and gadgets that gathers and shares data. AII concentrates on the clever, undetectable integration of such technology, whereas IoT offers connectivity.

9. **Harvesting Energy**

   the procedure for gathering and storing energy from outside sources, including solar radiation or vibrations in the surrounding environment, in order to power sensors and other equipment within an AII ecosystem.

10. **Privacy and Security of Data**

    steps taken to guard against misuse, breaches, and illegal access to sensitive data inside an AII system.

includes adherence to legal frameworks, anonymization, and encryption.

11. **The use of predictive analytics**
the process of forecasting future patterns or behaviors using data, algorithms, and machine learning. Predictive analytics is useful in AII applications such as healthcare monitoring and traffic management.

12. **Autonomous Systems** Systems that can carry out operations without the assistance of a human. Autonomous vehicles and AII-integrated robotic warehouse systems are two examples.

13. **Environmental Footprint** The effects of AII technologies on the environment as determined by resource usage, waste generation, and energy consumption. To reduce this impact, AII promotes sustainable practices.

14. In order to ensure justice, accountability, and transparency in AII applications, ethical AI refers to

the responsible creation and implementation of AI systems.

15. **Traceability** The capacity to follow an item's location, history, or usage using recorded information. By incorporating invisible monitoring systems, AII improves traceability in sectors like retail and supply chains.

16. This glossary ensures that readers have a clear and consistent grasp of Invisible Intelligence: Smart Sensing Transforming Everyday Life by providing a guide to the specific terminology used throughout the book.

ABOUT THE AUTHOR

Author and thought leader in the IT field Taylor Royce is well known. He has a two-decade career and is an expert at tech trend analysis and forecasting, which enables a wide audience to understand complicated concepts.

Royce's considerable involvement in the IT industry stemmed from his passion with technology, which he developed during his computer science studies. He has extensive knowledge of the industry because of his experience in both software development and strategic consulting.

Known for his research and lucidity, he has written multiple best-selling books and contributed to esteemed tech periodicals. Translations of Royce's books throughout the world demonstrate his impact.

Royce is a well-known authority on emerging technologies and their effects on society, frequently requested as a speaker at international conferences and as a guest on tech podcasts. He promotes the development of ethical technology, emphasizing problems like data privacy and the digital divide.

In addition, with a focus on sustainable industry growth, Royce mentors upcoming tech experts and supports IT education projects. Taylor Royce is well known for his ability to combine analytical thinking with technical know-how. He sees a time when technology will ethically benefit humanity.

www.ingramcontent.com/pod-product-compliance
Lightning Source LLC
Chambersburg PA
CBHW071055240526
45469CB00006BD/2304